YOUNG JOHNNY HARBAUGH

COLLECTED DESIGNS AND RENDERINGS

By DANIEL WARVELLE HARBAUGH

Young Johnny Harbaugh

Danzmark Productions, Houston, Texas

ISBN 978-1-304-97624-6

9 781304 976246

EDITOR'S FOREWORD

This is a compilation of the designs and renderings of young Johnny Harbaugh in the tremulous years prior to World War II. Many to most of these military related works were seized by the U.S. Government to prevent the advanced and vital military information to the enemy. They have now been declassified and available for public viewing. In the post- war era, John progressed on to being a Professor Emeritus at Stanford University, but his work presented in this book should inspire generations to come.

Young Johnny Harbaugh

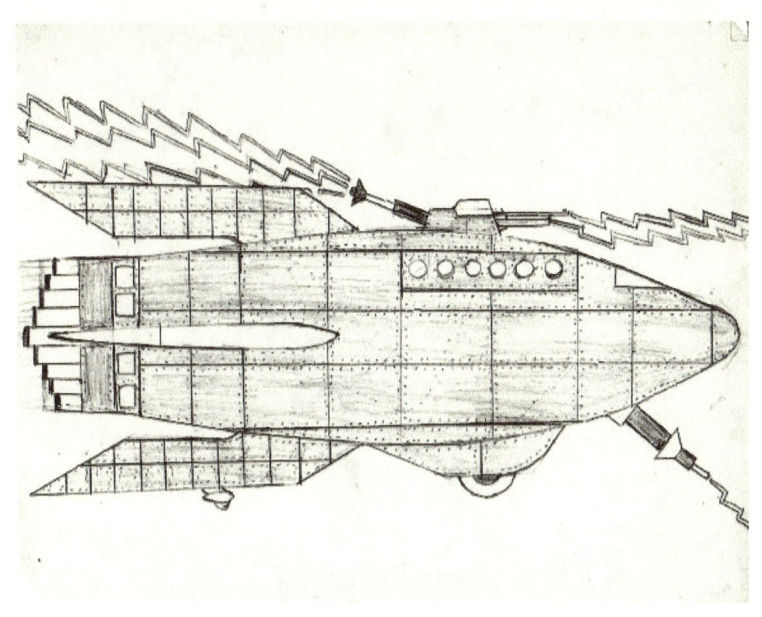

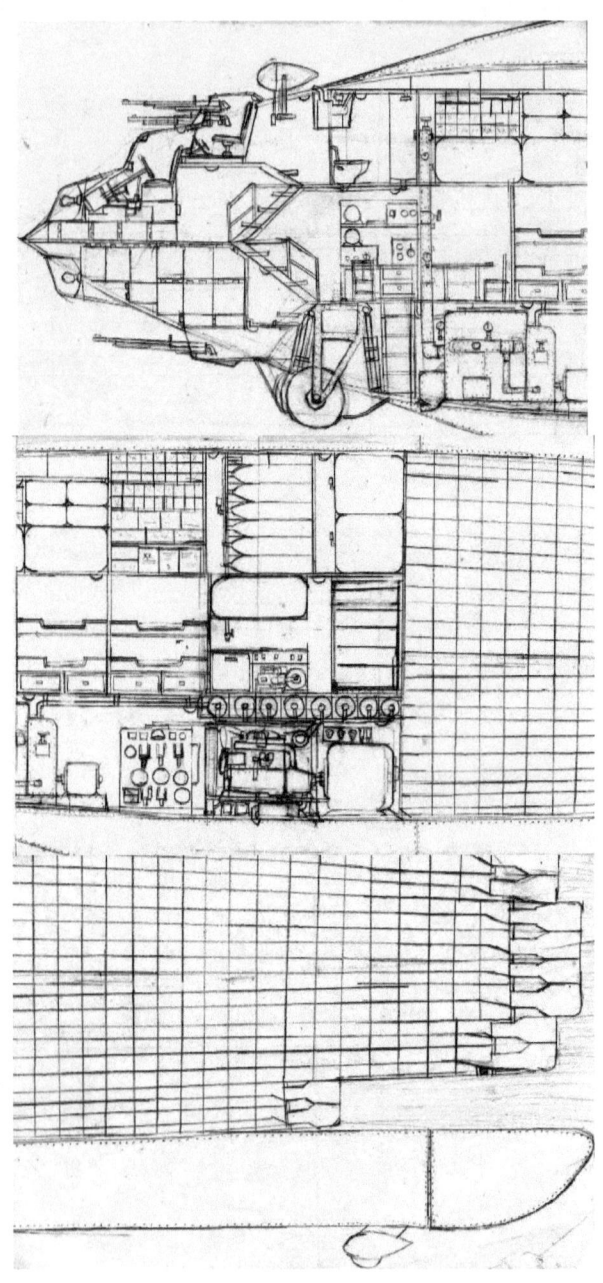

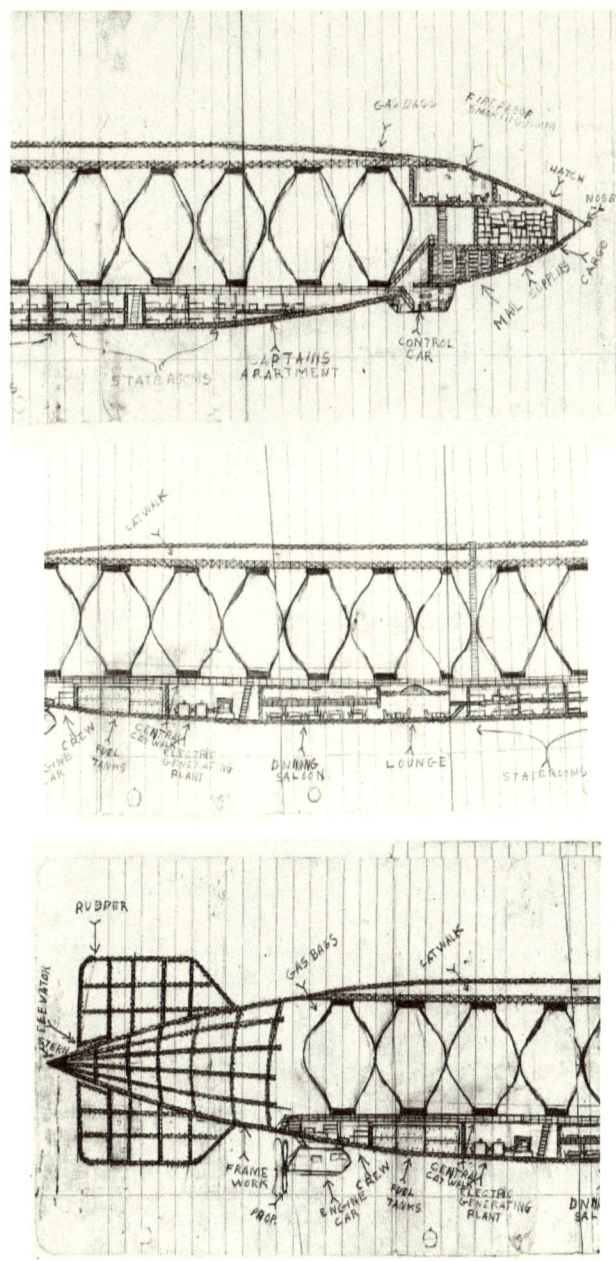

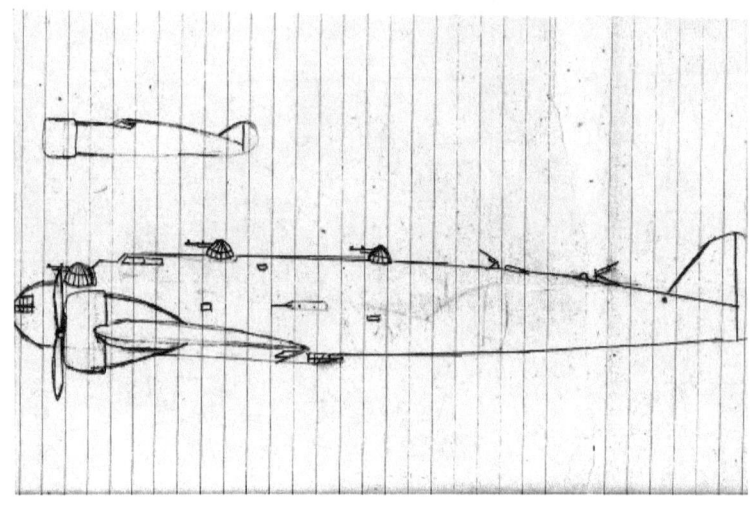

Young Johnny Harbaugh

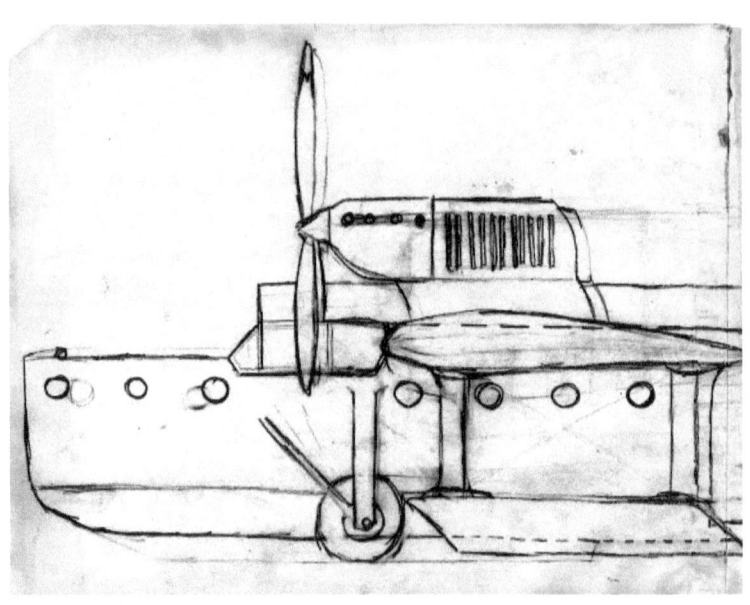

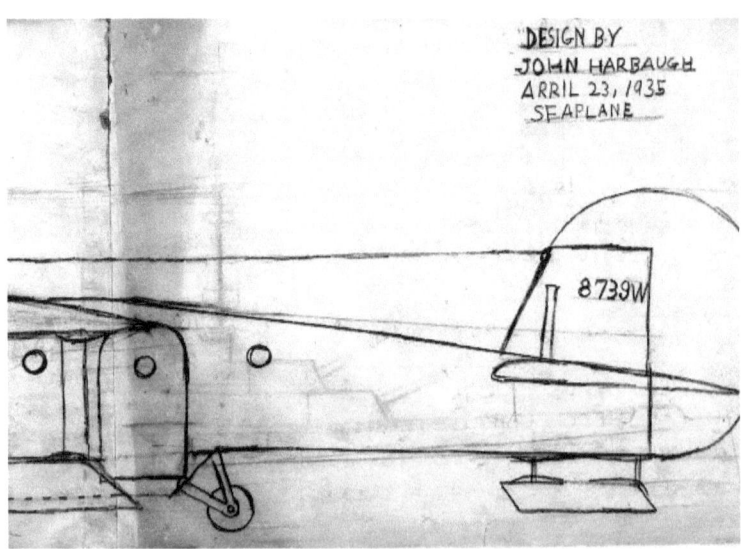

DESIGN BY
JOHN HARBAUGH
ARRIL 23, 1935
SEAPLANE

8739W

Young Johnny Harbaugh

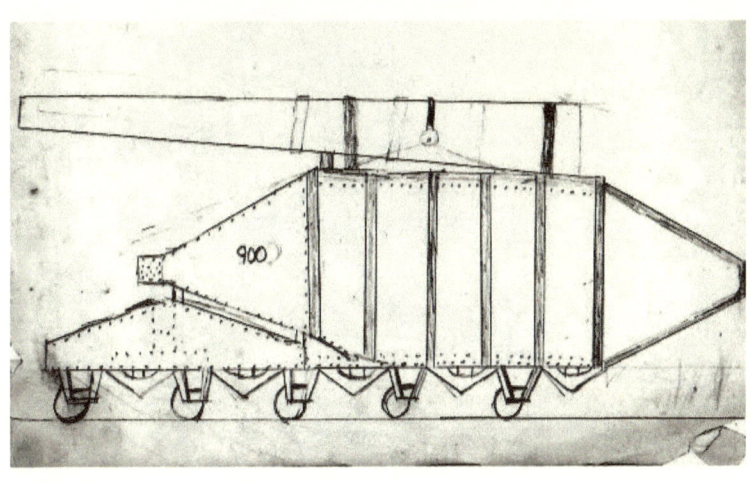

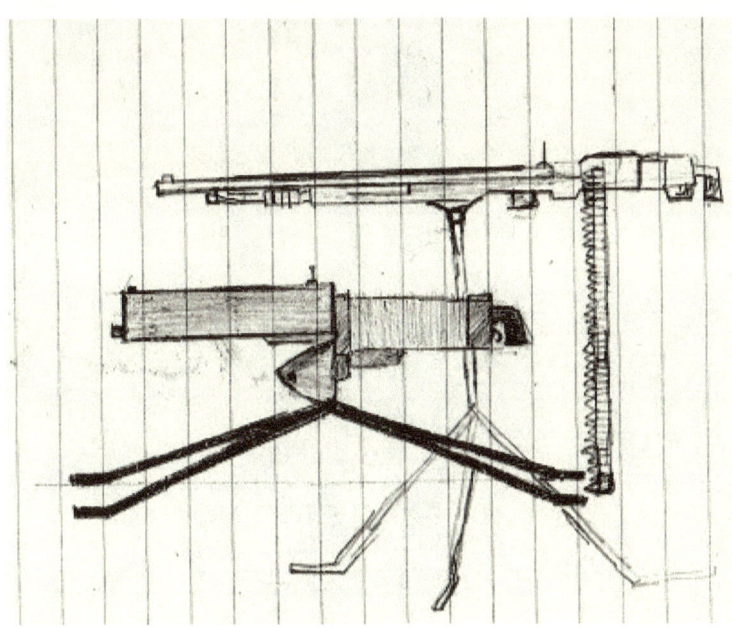

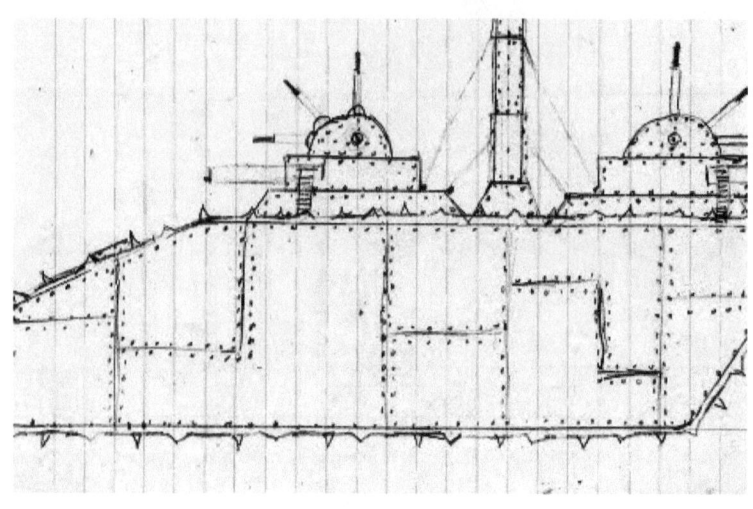

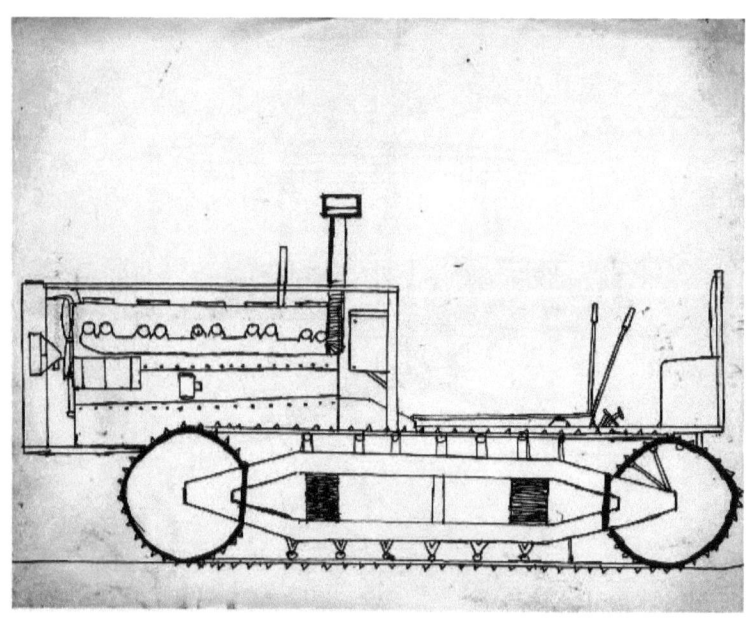

Young Johnny Harbaugh

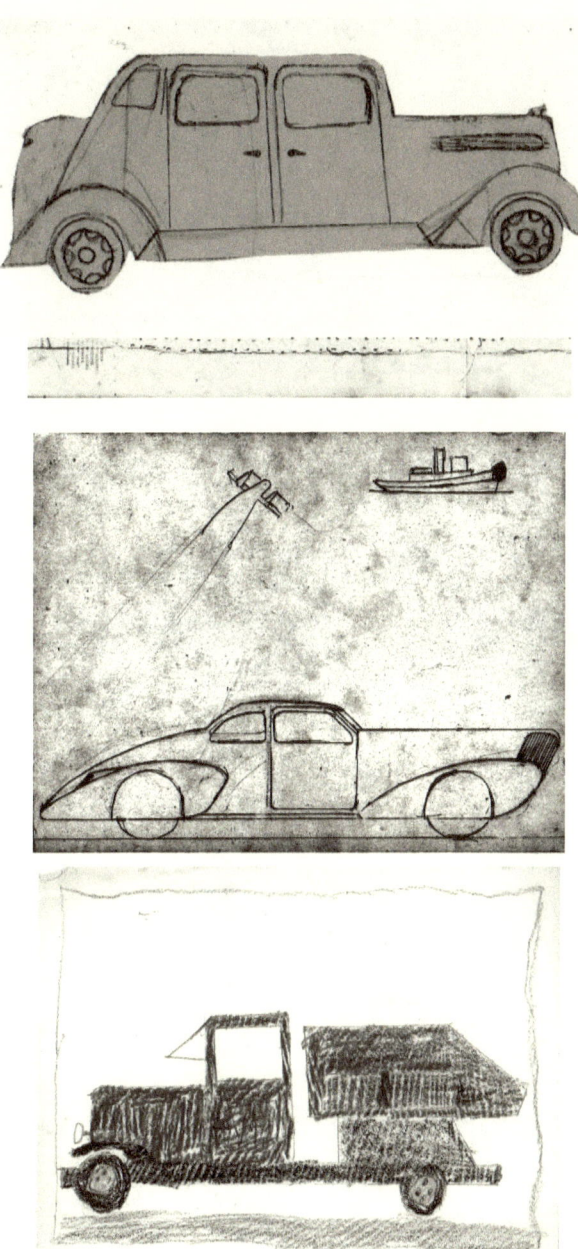

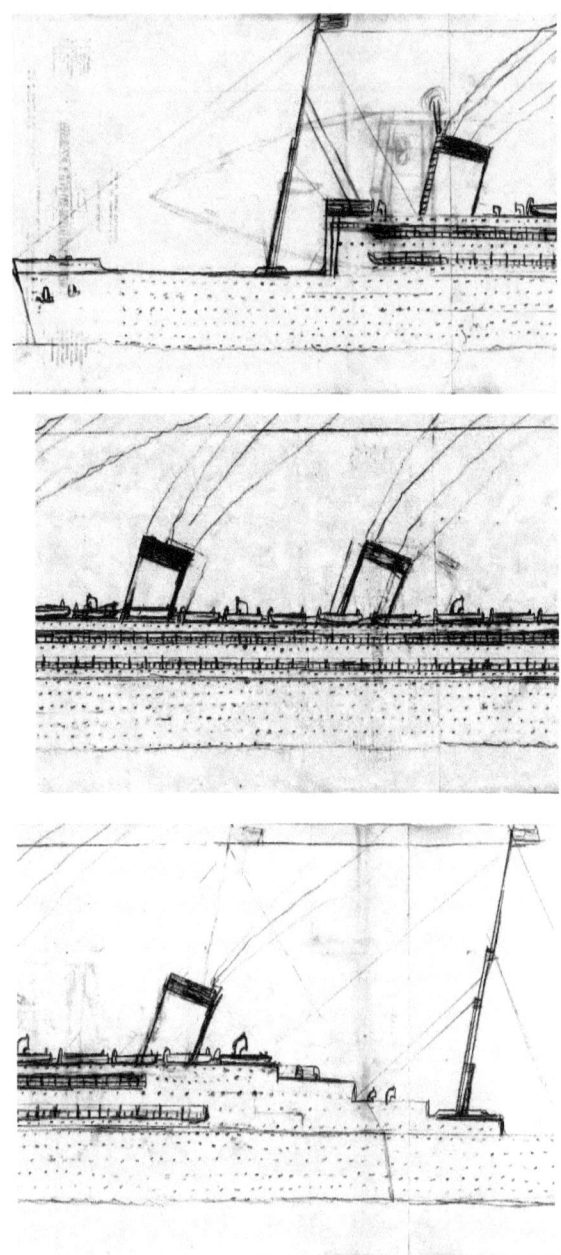

Young Johnny Harbaugh

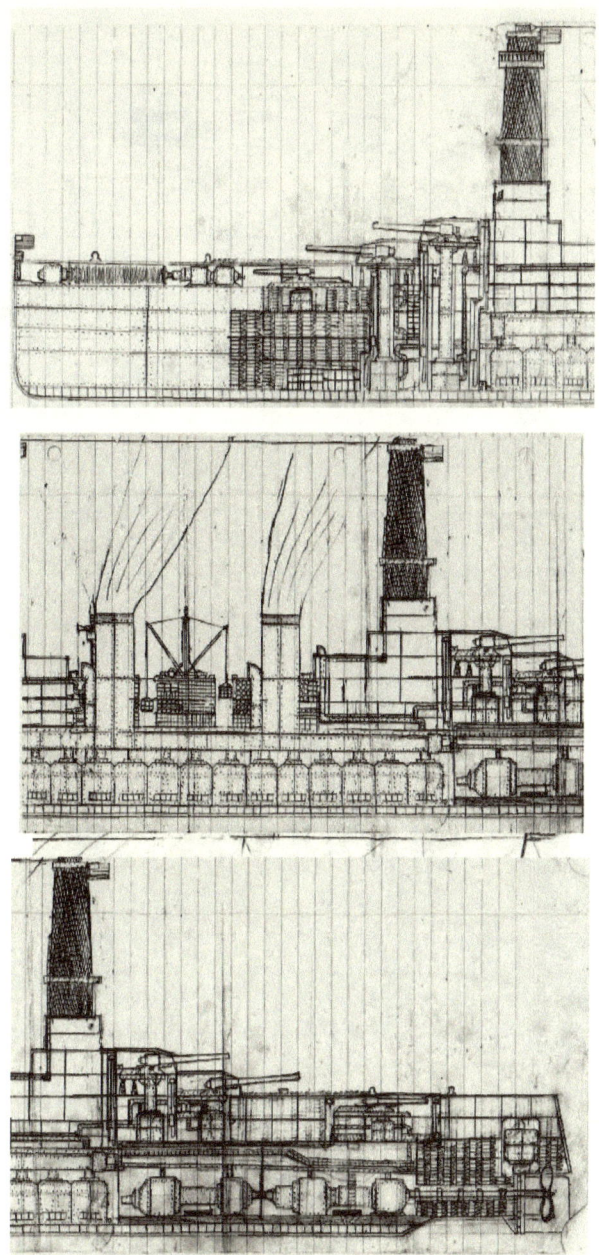

Young Johnny Harbaugh

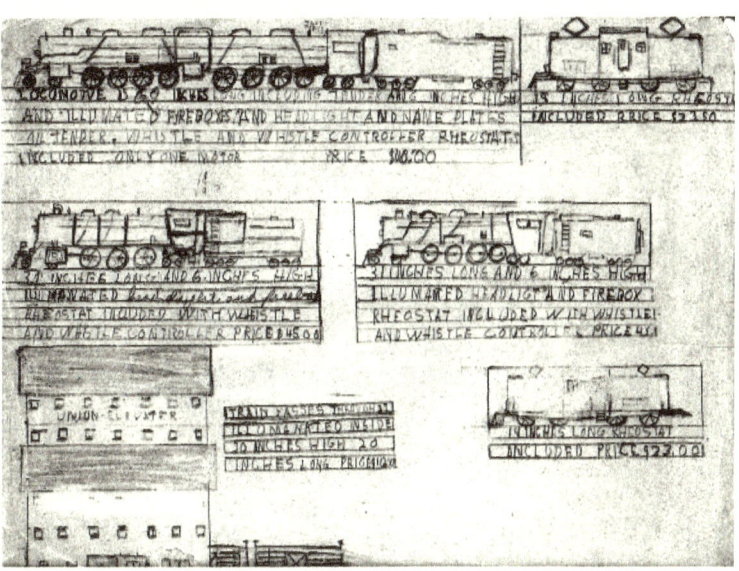

Young Johnny Harbaugh

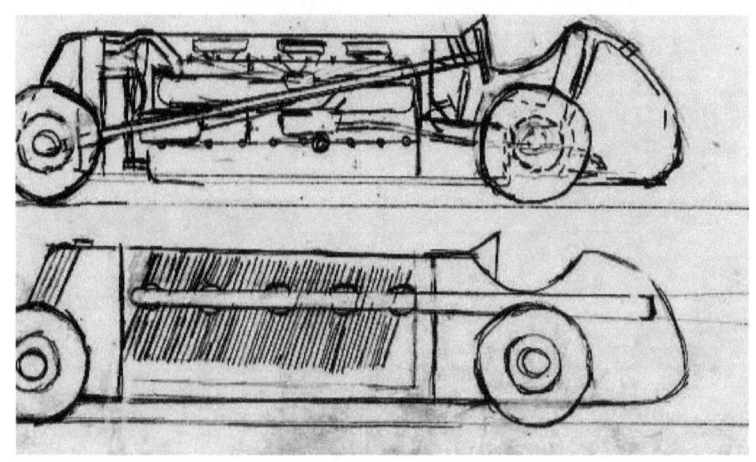

Preparing for his later career at Stanford, John
diligently utilized his Stanford Speller.

Young Johnny Harbaugh

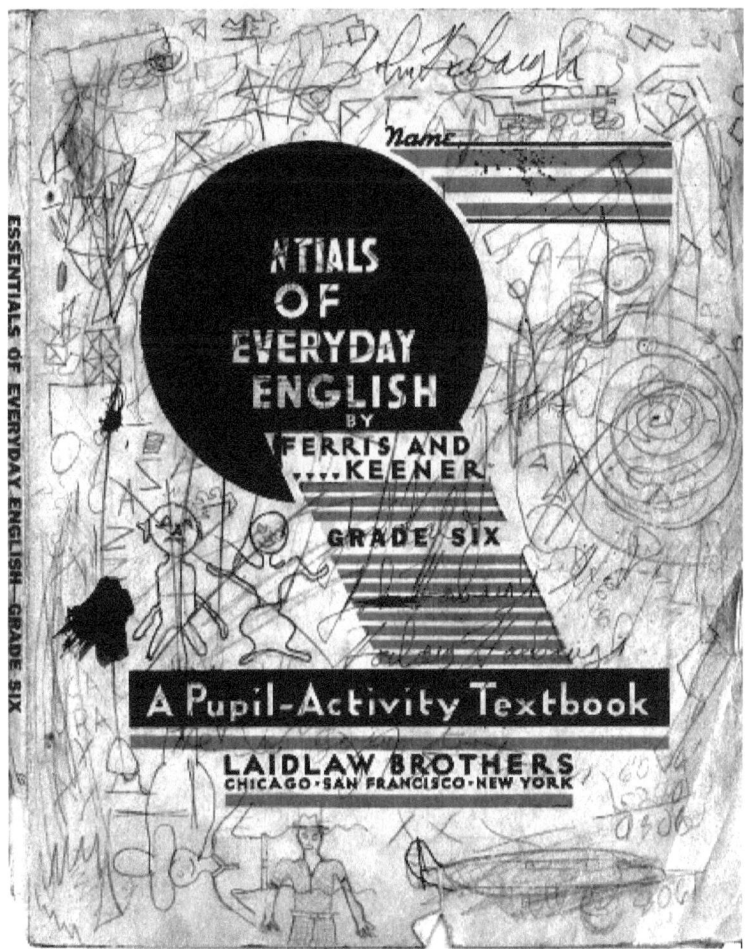

Johnny was a keen student of everyday English.

Young Johnny Harbaugh

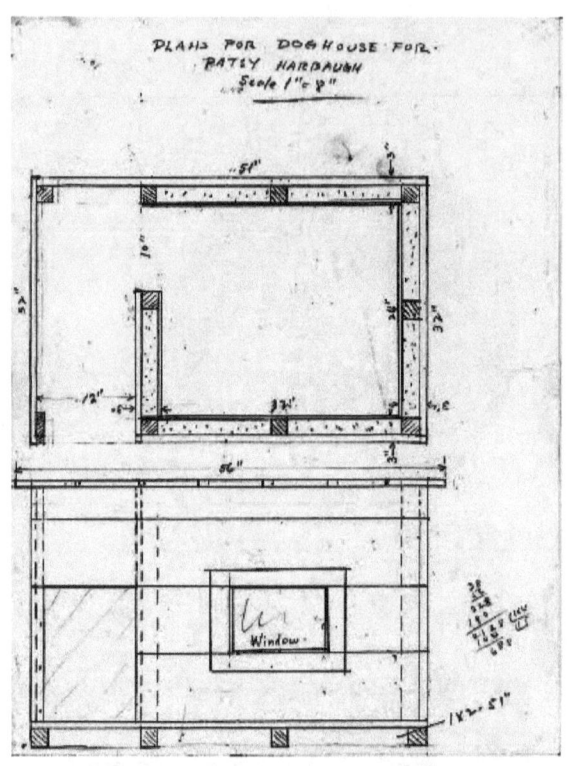

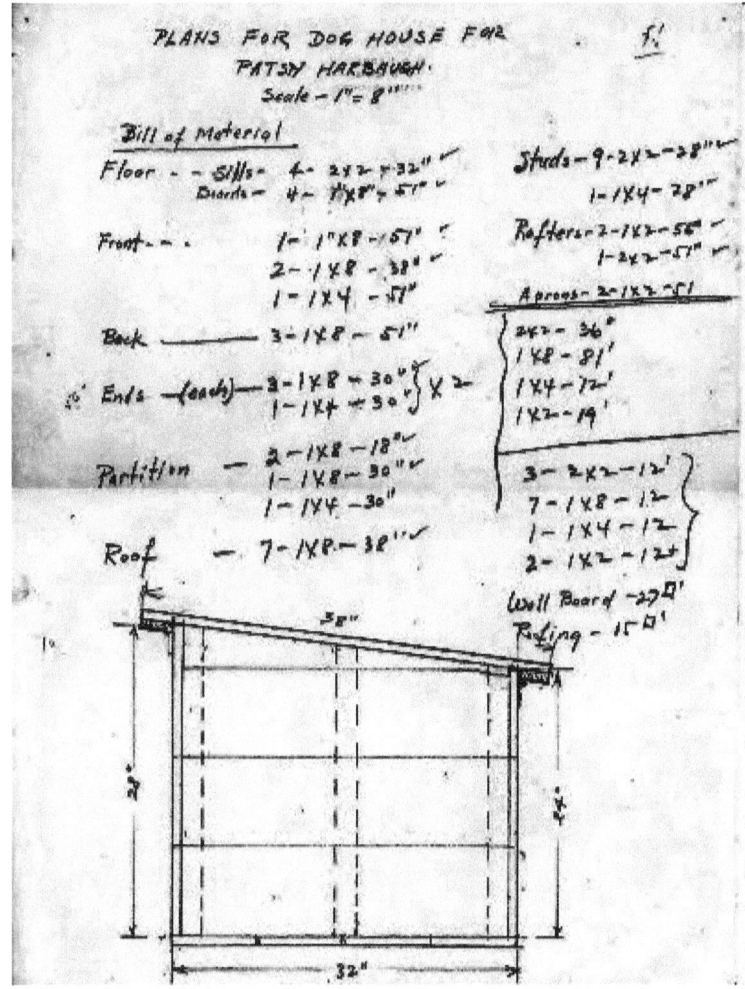

Dog House plans for 'Patsy' .

Young Johnny Harbaugh

Young Johnny Harbaugh

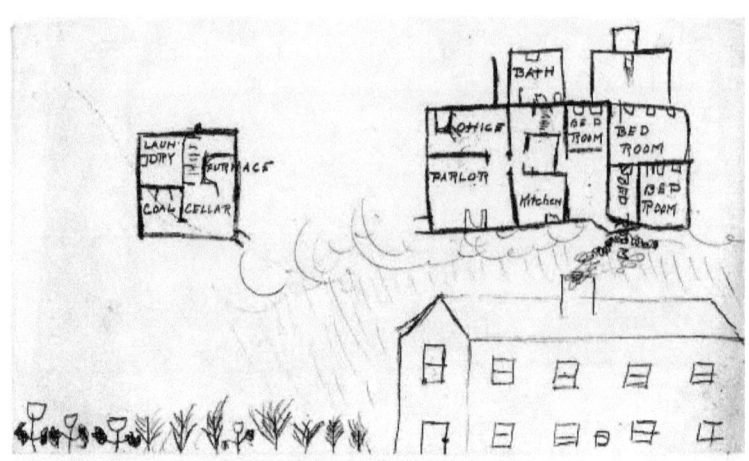

Young Johnny Harbaugh

Young Johnny Harbaugh

Young Johnny Harbaugh

The original painting was in water-color.

Always concerned with the public health,
Johnny created this good advice poster.

Young Johnny Harbaugh

88 years of progress - August 6, 1926 to 2014

Young Johnny Harbaugh

Young Johnny Harbaugh